Just a Moment

Titshall photographs of working lives

David Kindred and Roger Smith

Old Pond
PUBLISHING LTD

*The title page picture of men on the stack
is believed to show farmer John George (in the Trilby hat)
and workers at Wetherden.*

Published by
Old Pond Publishing Ltd
Dencora Business Centre
36 White House Road
Ipswich
IP1 5LT
United Kingdom

www.oldpond.com

Cover design and book layout by Liz Whatling
Printed in Malta on behalf of Compass Press Limited

Contents

INTRODUCTION

This is the second collection of photographs by the Titshall brothers. In the introduction to the first book – *In a Long Day* – I described how Ipswich man Doug Cotton saved the glass negatives from a leaking garden shed and lovingly salvaged as many as he could.

Doug passed them on to me in the knowledge that I was interested in vintage images and would preserve and publish as many as possible.

A half-plate camera. It was a camera similar to this that the Titshall Brothers used in 1925-35 when they took the photographs in this collection. Leonard Titshall would have placed the camera on a tripod. He covered his head and the back of the camera with a black cloth to keep out the light so that he could see the image on a ground glass screen. He could then adjust focus. This camera design was first used by Victorian photographers and was still used professionally by many until the 1940s.

The half-plate negatives measured 6½ by 4¾ inches. Their condition was very mixed. Some were almost perfect, but in the main they needed washing before printing. At the end of the 1980s I was still working with the traditional darkroom techniques which had changed very little since the Titshall Brothers were taking the photographs decades earlier. I spent hundreds of hours on the project in order to rescue as many of the images as possible.

Each photograph then had to be contact-printed as enlarging equipment for half-plate negatives was obsolete. This made shading exposure difficult and meant that many of the images were beyond printing. I did manage to print around half the plates and in 1999 *In a Long Day* was published, much to the delight of Doug Cotton.

The typical condition of a Titshall plate before cropping and restoration. The new image is shown on page 27.

Digital techniques have now largely replaced darkroom work. For this second collection I have been able to use a scanner to digitise the large glass negatives. I have used Photoshop software to 'repair' areas where the negatives were broken or scratched, or in some cases where the image was damaged by damp. Unfortunately, where damage to the edges was severe I have had to produce a cropped version.

Doug Cotton died at Christmas time 2005. I am proud that he trusted me to look after his cherished negatives and that we have been able to publish a second record of the 'Old Suffolk' that he loved.

Doug Cotton at the controls of his traction engine at Barham in May 1967. The engine was built by Ransomes Sims and Jefferies, Ipswich, in 1909 and was used to shunt railway wagons at the Ransomes Orwell works. Doug Cotton worked for Ransomes and was a regular visitor at steam rallies with his machine.

David Kindred
2007

Farm Cultivations

As we indicated in the companion book to this, *In a Long Day*, we still do not know for certain why Leonard Titshall motorcycled out from Ipswich to take these rural photographs. The assumption is that he 'cold-called', asking the workmen to pause for a moment or two so that he could take their portraits. Back at their Spring Road studio his brother Ralph made the prints for which the subjects paid a few shillings. Many local homes have a 'Titshall' up on the wall that is likely to have been printed soon after the photograph was taken.

Where pictures have been identified, they all seem to fit into the 1925-35 period, with a peak around 1929-32. It may be no coincidence that these were the years when three of Britain's most successful books about farming life were first published: H W Freeman's *Joseph and His Brethren* (1929), Adrian Bell's *Corduroy* (1930) and A G Street's *Farmer's Glory* (1932).

The Titshall photographs add an extra dimension to these popular books, showing in pictures the working lives that the books' authors were capturing in print.

Quotations in the captions are edited responses received by the *East Anglian Daily Times* and the Ipswich *Evening Star* when photographs were published in those newspapers.

1. A pair of Suffolks working with a wooden-beamed plough that looks as though it has seen better days. This would have been strenuous work on heavy boulder clay.

2. Summer with the Shire (front) and Shire-cross horses swishing their tails against the flies. The protective covering would have been half a pulp bag. The plough's right-hand wooden handle (from the ploughman's position) is an accessory. Plate 10 in *In a Long Day* shows a similar plough being used with just one handle.

3. Bait time: bread, cheese and an onion, with probably cold tea to drink. The ploughman is holding a pen-knife in his right hand.

4. A pair of excellent Suffolks, a young one in front and an older one behind, ploughing towards the end of the summer. The plough is probably a Ransomes YL.

5. Ploughing previously cultivated land. The horse at the back is decorated with a face brass 'to keep away the devil'. The ploughman is wearing the traditional corduroy waistcoat with heavy worsted sleeves. His lower legs are protected by buskins.

6. A pair of Suffolks turning on the headland.

7. There is no wheel on the plough. This saved money when the plough was bought but made the work harder. The horses are Suffolk-types.

8. From front to back: a Percheron, Shire-cross and Suffolk. They are breaking up a pasture with a double-furrow plough. Judging by its long hooves it looks as though the Percheron is ready for the blacksmith.

9. Fields were ploughed in strips known as stetches. With a single-furrow plough the ploughman walked about eleven miles to complete an acre. In this case, on sticky soil.

10. Here the pattern of stetches shows up clearly in the background. The ploughman has apparently cut off the end of the plough breast to make ploughing easier on this heavy land. This leaves the ground more open to be broken down by winter frosts.

11. A nice open winter's day in which to enjoy working in the sunshine. The ploughman's clay pipe is deliberately short-stemmed so that if he is knocked by the horse the stem will do less damage to his throat. The horses' square blinkers are the classic East Anglian type.

12. A three-horse team pulling a two-furrow plough. The horses are blanketed with pulp sacks to protect them from the summertime flies.

13. April or May, ploughing on summerland with a fine, well-groomed pair of Suffolks. 'My mother, aged 86, recognises the ploughman to be her late father-in-law, David Wyartt, with his daughter Mary. He was my great-grandfather and was working for Sir Gerald Ryan at Chattisham Hall Farm.' Christopher Wyartt.

14. A ploughman showing pride in his work with a well-matched pair of horses nicely turned out for work with brasses and ribbons. He is ploughing deep and coming up to the headland.

15. Late summer: three two-horse teams of cross-breds and Suffolks laying furrows over ready for drilling –
perhaps winter wheat.

16. Three-quarter bred Shires making baulks or ridges in preparation either for planting potatoes or sowing turnips.

17. A Suffolk (behind) and three-quarter Suffolk pulling a potato spinner which lifted the crop out of the ground ready for teams (usually of women and children) to pick them up by hand. Some people recall childhoods in which they helped out on the farm rather than go to school right from the beginning of harvest to the end of the potato lifting.

18. Good horses (these three are Shire-types) were needed to work the drill. This looks to be a Smyth model from the 1930s with which one man could both steer the machine and guide the horses.

19. The rough coat on the Percheron (nearest the camera) indicates that this is wintertime. The more usual team of one man operating the drill and one driving the horses are working on a nice fine seedbed. On the right of the picture the general store prominently advertises Colman's starch.

20. Again, a Smyth-type drill. The Shire-type horses are probably drilling stubble turnips for sheep.

21. Six generations of the Watkins family have lived at Gate Farm, Flowton, since Robert Watkins bought it in 1894. At the time of this photograph, 1926, the farm was being run by Robert's son Charles whose own sons were using a nice pair of Suffolks to drill winter wheat after a crop of beans in November. Cliff was at the back of the drill, Percy steering at the front, both in hob-nail boots and buskins.

22. A Suffolk and two Shires working with a Ransomes cultivator, probably in early spring, breaking up the ground ready for sowing. The background shows well-kept hedges and fields.

23. Two Suffolks and a Shire with a cultivator. The lead horse is in the middle – identified by the lead rein hanging down from the harness.

24. The heavy duck-foot harrow with shallow, wide tines is probably being used to get the ground ready for a winter crop. The strong wooden handles are not for steering – they are for lifting the harrow so that it can be cleaned. There are three horses, one of which is mostly obscured. The middle one might have been a young horse that the man at the back is driving. Behind him a Cambridge roll is working.

25. The Suffolk-type horse in the shafts is steering the roll and keeping it balanced while beside it a younger horse is probably being broken in to work. The Cambridge roll was usually used to press in seeds after sowing though it also often made corrugations before seeding. It's wintertime with a newly laid hedge in the background and pollarded trees.

26. Heavily built, Shire-type horses and factory-made zigzag harrows, harrowing-in a seedbed. Judging by the stacks in the background it's a reasonable-sized farm. Chickens are taking advantage of the newly sown ground.

27. Mixed pair of horses working a heavy harrow with the stackyard behind them.

28. The horseman is carrying a stick, perhaps to encourage the horse nearest the camera to pull its weight with the heavy harrows known as cromes.

29. Heavy harrows, or cromes, which were adjustable so that they could be set deep or shallow. The handles were for pulling up to clean. Working on light-ish land, the horseman has his coat draped over the horse's hames – perhaps anticipating a light shower or cold wind.

30. Two Suffolks and two Shire-types sweating with the effort of pulling the heavy crab harrow across clods to break them down. A large team of this sort had to be well trained so that they responded together at word of mouth.

31. Well-bred Suffolks pulling a Ransomes spring-tine cultivator on heavy clay.

Other Farming

We do not believe that the Titshalls set out to make a complete chronicle of farming. For instance, we know of very few photographs taken by them of the grain harvest - perhaps the most important event of the Suffolk farming year. Although many farms in East Anglia had cattle in the 1920s, the Titshalls seem to have taken very few livestock photographs. There is a higher proportion of winter photographs than might be expected. Perhaps the Titshalls spent the summer photographing weddings and other social occasions. The negatives of this type of photograph have not survived.

The pictures in this section record what was passing and what was to come. Traditional clothing, traditional horsemanship and well-proven working practices on the one hand and the lads on their tractors on the other. What must horsemen have made of the fellow whose tractor wheels are covered in soil in Plate 65? The arguments over how to care for the land and leave it better than you found it were just about to get into full swing.

32. Four cross-bred horses probably being brought home after work (judging by the mud on the road).

33. A group of Suffolk-type horses turned out into a well-strawed yard that was traditional in Suffolk. Part of it would be covered to give protection from the elements.

34. The shadows are long, but the workman, horseman and horses look fresh, so this is probably first thing in the morning. The horses usually went out at about 7-7.30 a.m.

35. 'The photograph was taken between 1927 and 1931 at the blacksmith's shop in College Road, Framlingham. The man standing by the horse's head is my late father Tom Card. Mr Fairhead, is the other man. It is now used as a technology workshop. I was born in the house next to the shop.' Ray Card.

36. The traditional braiding on this Suffolk stallion was probably to show off the crest of the neck. Perhaps the well-handled, well-behaved animal was being prepared for a show.

37. Was this light horse owned by a publican, as suggested by the benches outside the building?

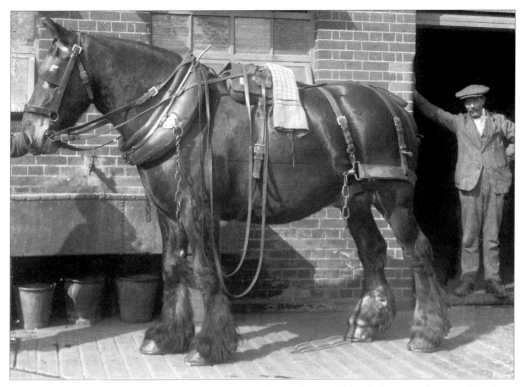

38. A Shire-type, of older conformation, this massive gelding is harnessed for shaft work – for general haulage or perhaps for brewery work. Its shoes have been strengthened for heavy road wear.

39. The gelding, suitable for light work, looks as though he has just been watered in the farm pond.

40. Harvesting with a Suffolk wagon which probably had a quarter lock. The design goes back to the mid-
nineteenth century. The same group of people are to be seen loading a different wagon in *In a Long Day*,
Plate 39.

41. Jack Jay has identified this harvest scene with a boat wagon It is at Mill Farm on the now built-up eastern
edge of Ipswich. The man on the left of the picture is Jack's father, Arthur Jay. The man to the right of the
picture is believed to be Len Rowe. The lad sitting on the horse, and the girl, were members of the
Stannard family, relatives of Mr Rowe.

42. Teams of harvesters with harvest ladders up on their carts.

43. The wording on the artillery wheel hub reads Woods & Co Stowmarket. Woods were iron-founders who manufactured a range of carts that were popular in Suffolk. Perhaps the coomb sacks are filled with barley grain threshed at harvest-time for the maltsters.

44. A light harvest wagon with full lock, wooden hubs and iron axles, perhaps loaded with chaff from threshing.

45. The level crossing at Mill Lane, Baylham where, Heather Excell writes, ' I have spent many hours for a lot of years waiting with the cows from Rodwell Farm, taking them back to the meadows after milking. The man in the bowler hat and carrying the *East Anglian Daily Times* is a farmer by the name of Joey Laws. He came from Moat Farm, Baylham, and apparently wore a bowler all the time.' See also Plate 160.

46. Mixed breeds of cattle, including Friesians and shorthorns, refreshing themselves on a warm day.

47. When this photograph of the red poll bull in show condition was published in the *East Anglian Daily Times*, Jennifer Johnston responded: 'Although taken several years before I was born, I recognise the two men standing on the right as brothers Cecil and Clarence Warne, my late father Cecil being second from the right. I believe the farm to be at Henley, although I am unsure of its name. The gentleman on the left of the photograph is possibly the farmer, a Mr Morton.'

48. Jack Amos (1903-97) who lived in a cottage at Hill House farm, Bucklesham. As shown in the frontispiece to *In a Long Day*, Jack had probably been using his hoe to chop out the remnant of a root crop for the ewes. Netting – in Suffolk often preferred to the heavier hurdles – separates the sheep and lambs from the ground that is being ploughed.

49. Lt. Colonel Cyril Alexander Barron (right) of Swilland with a two-wheeled miller's cart and three of his staff. Colonel Barron operated the wind and steam mill at Swilland. See *In a Long Day*, Plate 153.

50. Pollarded willows provide the background to Norfolk bronze turkeys.

51. This is probably a temporary clamp being set up to store sugar beet until it was collected by the haulier for delivery to the factory. The well-trained horses have been backed up to the clamp. Tipping the tumbrel was not an option because the haulier required his load close to the lorry and not spread over a wide area. The farm worker would remove the back board and clear a space to stand. He would then fork his way forward throwing the beet high and far behind him. Then it was back to the field for another load.

52. The sugar beet harvest. The wagon in the background is probably the back of a lorry obscured by the cart horse.

53. The front (trace) horse has road harness whereas its companion looks like a farm horse. It is possible that the trace horse and driver have been hired to help with a job of carting such as taking beet to the factory. Behind them is a well-used, muddy road and a factory-style building.

54. The horses are carrying plough trace. Perhaps the cart is carrying a set of heavy harrows in the cart being taken for work.

55. Ronald Tucker (1919-2005) left school at fourteen to become a cowman for Pauls & Son at Freston. He spent his whole working life with Paul's during the course of which his red poll cattle won many prizes at local shows. Here he is carting mangolds for cattle feed. It is believed to be his first day of work in 1932.

56. Mangolds being carted. The boiler suit was something of an innovation for farm workers.

57. Taking mangolds from the clamp and loading them into a robust tip cart to be transported for feed. This required a quiet horse that would stand patiently.

58. 'It was taken at Hintlesham then known as Laurel bank, now as Hill Farm. I think the farm was owned or rented by Mr S Warth of Manor Farm, Hintlesham. The men in the picture left to right: Jim Welham, Ben Welham, Ned Weham (my father) and Taff Welham their father. I have a framed photo like this on my staircase.' D Welham.

59. The lightly sprung road wagon which would not necessarily have been used on the farm is carting kale. In the background is a haystack that has been cut into for winter feed. The man's Wellingtons were a novelty at this time.

60. The tip cart or tumbrel was of a design very similar to those manufactured at Stowmarket by Woods & Co. The bulges in the sacks suggest they might contain potatoes. The stacks have been thatched but not trimmed.

61. Harry Smith (on the left, wearing First World War puttees) and Robert Bedford worked at Red House Farm, Hintlesham where they were probably feeding kale to livestock. Both men lived in Duke Street. The farm was owned by Mr J Ladbroke.

62. Two pairs of horsemen going out to start work. Two of the horses have sacking tied over their dutfins at their noses for protection from flies. The horseman sitting astride has a 'frail basket' – a traditional container made of plaited straw for holding food.

63. The Fordson tractor was introduced to Britain in large numbers by the Ministry of Munitions for the ploughing campaigns at the end of the First World War. Later known as the Model 'F', this early American machine undercut most of the other tractors on the UK market. It was still being used to plough in traditional stetches.

64. A Fordson Model 'F' supplied by Ernest Stannard, a motor engineer from Eyke. The casting to the left of the manifold is a petrol tank for starting. The tractor was changed over to paraffin from the main tank once the engine was hot. It is pulling a two-furrow plough.

65. From 1919 to 1922 the Fordson Model 'F' was also manufactured in Cork, Ireland. Both the Irish machines and their American counterparts from Dearborn were supplied to Britain. The box on the side of the engine contains the trembler coils for the ignition. This machine is being used to plough-in stubble on wet ground.

66. The 'improved' Fordson, subsequently known as the 'N' or 'Standard' model, went into production at Cork in 1929. This tractor dates from the first year of production because the later mudguards were tapered at the rear to prevent damage from implements when turning. The driver is Kenny Nelson, working at the bottom of Freston hill, overlooking the river Orwell.

67. The International 10-20 was one of the most respected American tractors to be sold in Britain. It was very dependable. It had a smooth, quiet engine with the crankshaft running on roller bearings. Early models were sold in Britain under the McCormick-Deering name but this late example dates from 1928-30.

68. A break during threshing – the drum is just visible behind the ladder. The can probably contains beer or cold tea.

No doubt the Jack Russell in the foreground is looking forward to catching rats when the gang gets to the last of the stack.

69. A winter threshing scene in the stackyard at Belstead Hall, Ipswich, about 1929. The Burrell engine was owned by Mr A R Wilson of the hall. Driver Spencer Proctor is standing on the engine. The boy at his feet is Frank 'Pip' Southgate. The man in the middle with a fine moustache is Walter Grimwood. The Burrell was later allowed to fall into disrepair after it suffered frost damage. It was not preserved.

70. Threshing in warm weather, judging by the men's shirt sleeves. They are just taking the last of the crop from the stack at the front of the picture, feeding to a thresher hidden behind the cart. The portable engine is powering both the thresher and the elevator. The coomb sacks of threshed grain are being loaded into a double-shafted boat wagon ready to go on the road.

The Urban Horse

Road improvement from the mid-1700s had led to the development of lighter-bodied, better sprung horse-drawn vehicles of a wide range of types. After the 1840s rail gradually became the most popular choice for longer trips but horses still flourished for shorter journeys and for transport to and from the railways.

However, in the late 1920s horses were beginning to have to share their roads with motorised vehicles. Plate 82 bears witness to this with the Ford Model TT lined up with the more traditional Co-op delivery wagons. Some types of transport, such as cabs and buses, became mechanised very quickly. Others, such as the dairy float or the rag-and-bone man remained horse-drawn for longer. The jingle of harness and the sharp clop of hooves on the road were to continue to be familiar sounds until the 1950s.

71. The utilitarian sprung wagon with artillery wheels is working for the Ipswich sanitary authority. The horse has been clipped out to help keep it cool and clean.

72. Pony and trap with a blanket over the seat, perhaps to cover the driver's knees.

73. A well turned-out gig.

74. Probably late Victorian or early Edwardian estate houses. The man driving the governess car has a wing collar, a sign of respectability.

75. Freeman Stannard (known as Joe) raised ten children in Wickham Market with his wife, Mary. In the summer Joe did field work; in the winter he hunted rabbits, rats, mice and moles. He caught the rabbits for local farmer Roger Clough who paid 6s 6d per dozen, gutted and into crates. They were then sent by train to London.

The donkey lived by the back door in a roughly thatched brush shack with whinbush sides. Apparently it used to enjoy being fed coal by the boys. It was very tame. When Joe was shooting rabbits he would hide behind the donkey and use its back as a gun rest, the donkey being unmoved by the shots. The photograph was taken at the bottom of Wickham Market High Street at the junction with Spring Lane.

76. Greengrocer Mr Squirrel of Spring Road, Ipswich, on his rounds in the late 1920s with his pony Kitty and wagonette.

77. The Dog at Falkenham. These were licensed premises until the end of the 1970s. It is now a private residence. The local carrier was from Walton, near Felixstowe.

78. Charles Whinney (centre) of Rushmere traded as a general carter and contractor. These carts, on display at the Ipswich horse show, Alderman Road, were in regular use for hauling timber, bricks and sand. Charles's great-granddaughter recalls that he used to move sand and ballast from a pit on Playford Road.

79. Shire-type used to heavy work and turned out for the show with good, old-fashioned braiding.

80. Double-shafted wagon pulled by a pair of Suffolks.

81. Well-bred Suffolks promote Paul's animal feed.

82. Delivery staff from the Wickham Market Co-op line up outside the store around 1930. The motorised van is a Ford TT.

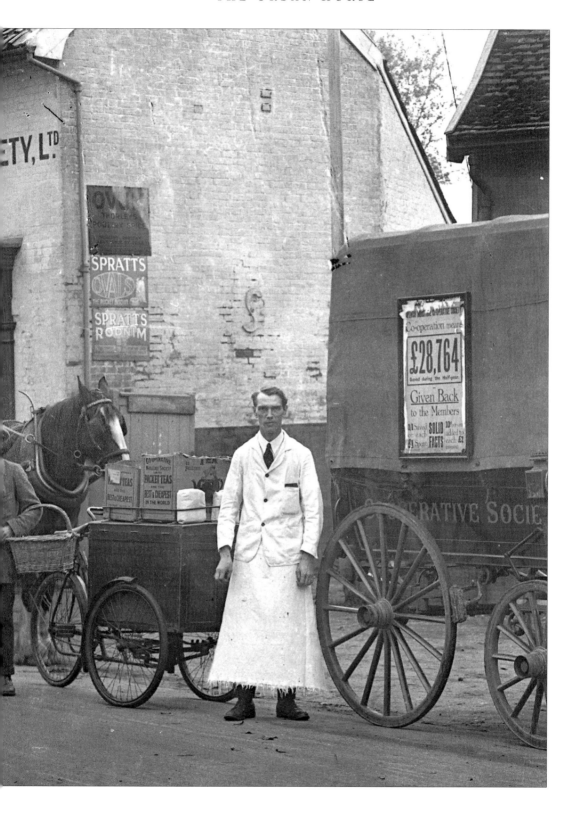

83. One of several vehicles owned by Lawn's of Wickham Market. See also Plate 109.

84. Smyth Bros were Ipswich-based ironmongers and gunsmiths with a showroom on the Cornhill and a shop in Fore Street. Second right is believed to be George Wicks who lived in Finchley Road, Ipswich.

85. The hardware store's brush wagon is in Lower Street, Stratford St Mary. The building with the poster is the Old Maltings which in the late 1930s was converted by Langham Fruit Farms for the cold storage of produce. It is now a furniture repository. 'It is said that in the First World War German prisoners were kept there.' Douglas Steward

86. Fore Street, Ipswich, around 1930. 'He was my father-in-law, Ernest Sadd. He worked for William Wakeling, a carter and removal contractor. Mr Sadd lived at the stables. The horse in the photograph was called "Captain". Bristo's garage now stands where William Wakeling had his premises at the top of the hill in Woodbridge Road, Ipswich.' Fred Maltby.

87. A sizeable tumbrel being used for dirty work.

88. The cart's backboard is supporting the shaft to reduce strain on the horse while bricks are being loaded.

89. A trolley wagon from Arthur Emeney, contractor, of High Street Hadleigh. Mr Emeny provided cartage for the London and North Eastern Railway from Hadleigh station.

90. The coal merchants AE Stow was founded in Ipswich in 1926. 'In wet weather the only protection the driver had was a coal sack, which had been cut to fit over his shoulders. From Monday to Friday ordered coal was delivered to customers but Saturdays were devoted to "hawking", the men selling sacks from the cart or lorry for cash. The horses covered the Nacton and Gainsborough estates whilst the lorries covered a wider and hillier part of the town.' Noel Stow (son of the founder).

91. The contractor and coal merchant with his sturdily built wagon.

92. Hauling pipes at Ipswich Dock.

Motorised Transport

In the late 1920s private motoring was in its infancy. There were high tariffs on imported cars, and although Ford began assembling its Model T at Manchester in 1911, it was not to build its Dagenham plant until 1932. Morris launched its first car available for under £100 in 1931. After that, motoring came to resemble what we recognise today, with a Highway Code from the same year and compulsory driving tests from 1935.

The first filling station had been built in 1920 by the AA as a service to its members. Most petrol suppliers were in small buildings like those shown in Plates 93-5. However, it was quite unusual for photographers to use these as their subjects at a time when the vogue was for the 'picturesque'. The Campaign for Rural England – founded in 1926 – stated at this time that 'to many people the petrol pump is the very symbol of ugliness.' Modern petrol station development was to begin in 1932 when Shell-Mex and BP launched a joint venture to build them.

In the meantime, someone with more moderate means could get on a motorbike like the BSA 557 cc side-valve model in the picture below. This was how Leonard Titshall roved the countryside for his pictures – with the camera and tripod in the sidecar.

93. Ruben Bryanton outside his splendid little service station at Witnesham in the 1920s. His arm is resting on a Bowser continuous hand pump. Behind the garage to the right is an ex-Great Eastern suburban railway carriage. The motorbike is a Triumph, probably a 500cc side-valve model from 1924-5.

94. The cottage is on the corner of Stratford Road and High Street, Dedham. The cottage is still there (although considerably more tidy) but the service station has gone. At the time of the photograph it probably sold fuel in cans. The motorbike is believed to be a 4-cylinder Indian dating from about 1925. The sidecar is a commercial outfit often used by tradesmen.

95. 'The man on the right, at a service station on the A140 (the main Ipswich-Norwich road), was my father George Banham. The service station was an old blacksmith's situated near the Magpie Inn. My grandfather Fred Banham and George started repairing and selling cycles and petrol. By the mid 1930s they moved to Springfield Garage, Stonham. The old service station was demolished in the late 1990s.' Jill Birkin

96. A flat-nose Morris Cowley.

97. A commercial vehicle has been built on to a pre-1924 Ford Model T car chassis.

98. The Citroen 5 CV, made in about 1923-4, is standing outside a nursing home.

99. The Model T has been modified with a special coach-built body to make it look more sporting than the usual model.

100. The Silver Queen bus service operated from Dovercourt. This Chevrolet is outside Harwich town hall. The registration is for Kesteven, Lincs, 1927 or so. Coach-builders in that area may have made the conversion that gave the vehicle a lengthened wheelbase and an extra rear axle to carry more weight.

101. 'Blue Bird' was the fleet name of Garnham's garage of Woodbridge. In 1934 it sold its bus services to the Eastern Counties Omnibus Company but continued to operate coaches on private hire work and school contracts. Garnham's closed in 1951. This service ran from the Ufford Crown to the Old Cattle Market in Ipswich. When the weather was poor the driver would make a detour to Lower Ufford to save children the walk to the Crown. The vehicle is a Ford TT with a commercial chassis and 1-ton payload.

102. The Tilling-Stevens petrol-electric bus had a maximum speed of twelve miles an hour on its service between Felixstowe and Stowmarket. Passengers on the top floor were apparently at some risk from the jettied upper stories of the older houses on some of the narrow Ipswich streets. Standing with the conductor is the terminus regulator, Mr Fussell, whose false hand was concealed in a glove. The photograph was taken in Ipswich outside the sorting office at the Old Cattle Market terminus.

103. An Ipswich trolleybus built by Garretts of Leiston, one of fifteen Garretts and fifteen Ransomes that were delivered in 1926 to replace the trams. The driver is standing at the new Adair Road, Bramford Road terminus. The destination indicator was a glass tube with different numbers around the outside. It could be rotated to show which service was running.

104. An outing from the Garland public house, Humber Doucy Lane, Ipswich, about to set off in a Dennis
charabanc belonging to the picturesquely named Marguerite company. It later amalgamated with
Primrose to become Primrose and Marguerite. The vehicle started life as a wartime 3-ton subsidy truck.
It was later re-bodied to take a charabanc.

105. Transferred from London to Ipswich in 1920, the Tilling-Stevens TS3 charabanc was operated by the
Eastern Counties Road Car company. They created the first bus network in Ipswich but also ran a
charabanc service. A petrol engine drove an electric dynamo so that an electric motor drove the back
axle. Having no gearbox or clutch, the vehicle was comparatively easy to drive.

Commercial Vehicles

One legacy of the First World War was a pool of ex-army vehicles and de-mobbed servicemen who knew how to drive them. At Slough, just west of London, there was a depot with acres of ex-army trucks for sale.

The Ford Model T, and its lengthened version the TT, became the basis for many different uses of vehicle, with coach-builders becoming increasingly skilled at improvising on Ford's chassis.

Throughout the Depression years of 1929-33 commercial motoring continued to grow. There was no stopping it once the business advantages of speed (compared with horses) and flexibility (compared with the railways) became clear. Although the national speed limit was restricted to 20 mph - the restriction was removed in 1930 for cars and motorbikes - the banana lorry in the photograph below could still get to Covent Garden, London, from Ipswich in four or five hours.

There were several banana merchants in the Ipswich area which had ripening rooms warmed by gas flames to retain humidity. This company later traded as J J Wilson (Ipswich) wholesale fruit merchants at 12 Tacket Street.

106. Albert Soames, who licensed this Ford Model TT from 1927 to 1932, was clearly adapting a horse-drawn trade to motor vehicles. 'The person standing at the front of the van at Clopton is my father George William Button who worked for Albert Soames senior. I think the other man is Albert Soames junior. The mobile shop later grew to become "Soames Forget-me-Not Coaches" now at Otley (formerly at Clopton). The slogan was Dad's idea, a clever wordplay during a period when other coach firms and the like were given names of flowers etc.' Brian Button.

107. Eighteen-year-old delivery driver D Smith poses with his gleaming Morris 'squashed bull-nose' 6-7 cwt van. A Yapp and Son, pastry cooks and confectioners, were based in Ipswich.

108. Many people in Ipswich still remember the delicious smell of coffee wafting across Westgate street from the Oriental Café at number 21. It was a popular venue for wedding receptions despite having a poltergeist. The café's delivery driver was able to park his Rover 8 hp twin-cylinder air-cooled van across London Road. This main route out of Ipswich towards Colchester now carries thousands of vehicles an hour.

109. Lawn's Stores was one of the great institutions of Wickham Market. There were two vans with drivers Cecil Boast and Percy Fryatt serving a large rural area including Easton, Letheringham, Charsfield, Hoo, Marlesford, the Glemhams, Pettistree, Ufford and Dallinghoo. Customers ordered groceries, provisions, tobacco, alcohol, hardware, paraffin, crockery and other necessaries one week for delivery the next. They were also local parcel carriers. The vehicle is a Morris-Commercial T-type, built around 1927.

110. The 7 cwt Trojan van, made in 1925-6, was used to fetch newspapers and provisions for the shop and newspaper round owned by Mrs V Willis at Crowe Hall cottages, Stutton. The man by the front wheel is Maurice Robinson. Various models of Trojan van became available in the 1930s, the 10 cwt one being the mainstay of the Brooke Bond Tea fleet. They had a chain drive to the rear wheels instead of a prop shaft. The solid tyres were narrow. David Barton recalled that 'the wheels were similar to the gauge of the tramlines in Ipswich. A driver wishing to turn from Tavern Street into Princes Street found himself going straight over the Cornhill!'

111. Ford Model TT.

112. The building of the milk-collection depot at Claydon was demolished when a flyover for the A14 was built. The Co-op collected milk in churns from farms in the surrounding district, pasteurised it and took bulk down to London at night.

113. E C Catton was a substantial business based at the Jubilee Bakery, Stoke, Ipswich, from which they used to serve the Shotley peninsula with their home-made produce.

114. This photograph can reasonably be dated to late 1929. The house in the background with its For Sale sign slightly obscured by a covering has been identified as the Mutton, Offton. In 1929, when it was known as Armigers, it was sold by a Mrs Symons to the parents of H W Freeman, author of *Joseph and his Brethren* and other Suffolk novels. One of Mrs Freeman's letters preserved at the Suffolk Record Office gives a completion date of October 26th. By 18th December they were in and distempering the house.

The Danish Bacon company was founded in 1902. Presumably the Danish bacon was brought across to East Anglia by sea on the Esjberg-Harwich route. The Gilford lorry, with its 6-cylinder side-valve engine, was considered fast for its day. It dates back to about 1928.

115. The Ford Model TT made in about 1925 is probably standing in front of Cann's shop window at the front of the dairy. A sign in the window advertises new-laid eggs for sale.

116. Well muffed-up against cold weather, the Ford Model TT was registered from 1927 to 1932 to Tyson and Son, a grocer of Woodbridge. This photograph was taken outside Tyson's garage in Little St John's Street. The lad holding what is probably lead piping is Jim Woodruffe.

117. Timber imported from the Baltic ports through Ipswich Dock involved haulage at several stages: from
dockside to sawmill; from sawmill to storage yard, and from storage yard to fabrication shop. This
shunting truck with a protective metal plate at the front was registered in 1930. Timber merchants
William Brown used whatever space was available on the dock as well as their own yards. The man
on the right has a leather shoulder strap for carrying the hard-edged load.

118. William Brown had several of these conversions which were a familiar sight around Ipswich: a Morris-
Commercial T-type customised to act as a builder's lorry with staging built up, crosspieces and leaf
springs bolted on to customised supports. Presumably the driver sat in the middle. The headlights have
been moved to the top of the cab which advertises Portland cement. Browns were closely associated with
the Claydon cement factory.

119. The truck with its heavy chassis and wooden wheels is probably an American import dating back to the war. Anglo-American was founded in 1888 to supply paraffin for lamps. In the early 1900s it began selling petrol imported from America under the name of Pratts Perfection Spirit. Anglo-American later became the Esso petroleum company (now Exxon).

The man climbing into the cab is believed to be William Buggs who worked all round the south-east with Anglo-American before settling in Ipswich in the 1920s. Pratts Spirit operated from Ipswich, either St Matthews Street or Chancery Road off Princes Street.

120. The Ford Model AA 30 cwt truck of about 1928 vintage is probably being operated by a local builder. Note how he has cushioned the load on top of his cab.

121. Ford Model TT 1-ton truck with disc wheels at the rear, built around 1925.

122. D. Quinton & Sons occupied nearly all the station yard in Needham Market, the vehicles being kept in former cart sheds where a parade of shops now stands. The firm brought in corn from farmers in about a seven-mile radius. Driver Charles Brett (right) had a mate to help load the sacks which could weigh up to 20 stones. Given an average of four loads per day, at eight tons per load, each man shifted about 16 tons a day on his back.

According to Charles Brett's son: 'probably the strangest load my father ever carried was lumps of chocolate. Early in the Second World War Norwich was bombed and a chocolate factory badly damaged. A large amount of chocolate was declared unfit for human consumption and consigned for pig feed. Two rail trucks of chocolate arrived at Needham at a time when the ration was 1 oz per week. It was not surprising that much of this failed to reach a pig trough!'

123. Coal merchants Thomas Moy were a national organisation whose name was to be seen at many railway goods yards. They had an office in Arcade Street, Ipswich. The truck is a Morris-Commercial T-type.

124. In the house-building boom of the late 1920s there was plenty of work for contractors around Ipswich – although not many can have come from as far afield as these. The vehicle is a Ford Model TT. The cushion tyres on the rear axle were designed to make driving softer than it was on solid tyres.

Men at Work

While village populations declined in the 1920s, towns such as Ipswich and Felixstowe grew rapidly. In Ipswich between 1921 and 1930 nearly 5,000 new houses were built, mostly as new homes but partly to replace slums. There was no JCB backhoe-loader to help with this work. It was mostly done by hand by gangs of labourers.

The Ipswich Dock – then the largest single dock in the country – was handling an increasing tonnage of grain, timber, coal and general merchandise. The net registered tonnage of vessels grew by nearly seventy per cent from 1920 to 1930. The impression is that, despite the agricultural depression, Ipswich was a thriving and growing town when the Titshalls took these photographs.

The railways, now the life-blood of land travel for longer journeys, employed large numbers of workmen. Those in the photograph below would probably have been railway employees from the district civil engineer's department. It is thought that this is Hadleigh Road railway bridge in Ipswich (see also Plate 126). When the picture appeared in the *East Anglian Daily Times* Grace Rochanski wrote that 'The railway bridge at the Hadleigh Road had just two tracks when I was a child. I can remember the works to extend the bridge around 1930.'

125. The J70 tram engine belonging to the London and North Eastern Railways was photographed at the Ipswich Cliff Quay. It was used to shunt trucks around the Dock area. The shunters with their poles are seen with the fireman (left) and Bob Fenning with the bowtie.

126. A view of Ipswich's Hadleigh Road bridge from close to the railway engineer's department. The left-hand arch was for the main East Suffolk line. The right-hand arch spanned the approach to the sidings where Manganese Bronze and one or two other companies ran their own engines.

127. This J15 class 0-6-0 engine was one of the mainstays of freight haulage in East Anglia. They were built from 1883 to 1913 but lasted right to the end of steam in this region – 1962. The driver and fireman are at the locomotive maintenance yard, Ipswich in the mid 1920s. The yard was close to the tunnel through Stoke Hill.

128. The F5 class steam locomotive 147E was at Hadleigh station in around 1930. On the side of the bunker it can be seen where the Great Eastern plate was taken off and replaced by a new LNER one after 1923. The man standing under the dome is fireman Frank Cocksedge. The Hadleigh branch from the main Ipswich-Colchester line opened in 1847, running Bentley-Capel-Raydon Wood-Hadleigh. It carried passengers until 1932 and goods until the outbreak of the Second World War when it was extensively used to supply the underground ammunition store in Hadleigh. It closed completely in 1965.

129. The railway fitting shop staff on a Class 1500 engine at Croft Street, Ipswich. Sitting on the left-hand bumper is Les Bloomfield. The man with his arm on the vacuum pipe is Sid Lincoln.

130. At Ipswich Locomotive yard in about 1923-28, a fine Claud Hamilton class engine. This was the class that pulled many of the mainline express trains in the 1920s and '30s. At the front is a stop board prohibiting others from buffering on when someone was working on the train.

131. Flint Wharf, Ipswich Dock, with the mills and silos of Cranfield Brothers and R & W Paul in the background. This area close to Stoke Bridge is currently being redeveloped.

132. The buoy-lifting raft is working in Ipswich Dock, south-west Quay, near the lock. The workmen lifted navigation buoys before cleaning, repairing and painting them. The raft was probably towed by a tug.

133. Flint Wharf, Ipswich Dock, with Tovell's Wharf in the background. 26 metres long and 6 metres across the beam the Thames barge *Repertor* (discoverer) was built in 1924, the first steel barge made by Horlocks of Mistley, Essex. She went on to trade general cargo up and down the east coast under sail until the mid-1970s. At the end of her working life she became a tanker barge. Now fully restored and available for charter from Maldon, Essex, she is a keen competitor in barge races, several times being Champion Barge.

134. The sailing barge *Petrel*, waiting to lock-out at Ipswich Dock. The barge was owned by millers Cranfields. Whereas these barges were usually light-laden going out, this appears to be heavily laden, presumably with flour.

135. The portable electric winch was used at Ipswich Dock for pulling ships in or to operate a crane – in this case to unload.

136. This is probably on the island site at the Ipswich Dock, next to the public warehouse. The wooden shovel to the left of the photograph suggests that the men might be filling the sacks with malt.

137. Probably a council lengthsman, the workman is wearing Wellington boots that were thought to be less comfortable than working boots. The fence posts behind him are typical of those erected after road straightening or widening.

138. Perhaps preparing for sewerage or a main storm drain.

139. Work interrupted by a fallen tree.

140. Cantilever trolleybus posts can be seen on the left and it is possible that the men are taking up the tramlines in Ipswich. The cobbles to the left of the picture had previously been laid between the tram rails and either side of them. The hoarding on the left is advertising the Empire cinema which was in the Social Settlement building in Fore Street. On the right of the picture is a gas lamp.

141. Concrete laying in St John's Road, Ipswich, where it cuts into Spring Road.

142. This is probably road building on one of the many new Ipswich estates. Temporary rail track has been laid to move materials and the concrete mixer around. The concrete mixer and crane were manufactured by Winget, founded in 1908 and still in the same industry today.

143. Perhaps the same scheme of works as in Plate 139.

144. Probably a quarry scene. On the top left-hand side is a puddle mill into which the workmen would fork limestone to be ground. It is powered by the portable engine.

145. Don Harman has recognised his father, George Harman, in the bottom right-hand corner of the tarmacadam mixer at the East Anglian Roadstone and Transport Company, New Cut East, Ipswich Dock. George Harman came to Ipswich in the 1920s and was superintendent at the plant until the '70s.

146. Tar men patching the road.

147. Aveling & Porter road roller No. 7058 was new to the Essex Steam Rolling Association in 1910. It was their fleet No. 22. In 1921 it was registered as No. 1981. It was last recorded in 1951.

148. The Showman's engine has John H Barker on the plate.

149. It is not known where the two fairground scenes on this page were photographed.
Ipswich did have a fairground site on London road.

At the Cottage Door

We have virtually no photographs taken by the Titshalls of the interiors of buildings. Away from their studio they needed a good source of daylight to prevent their exposures becoming impossibly long. They also needed space for the bulky camera and tripod.

They settled on a working practice that suited them and their clients, creating informal portraits against the background of their homes.

Some of these were also workplaces. The photograph of the beautiful sixteenth-century building on this page is of 31-33 High Street Debenham, where Mrs Carrie Aldous (in the shop window) ran her egg-collecting business. The man in the middle is Bill Aldous and the little girl is believed to be her grandchild. See also Plate 164.

150. 'The photo at Pettaugh is of my mother-in-law's mother, who was then Edna French (centre), who later married Stanley Kerry. She died in 1987. With her is her mother Henrietta French. The child on the left was an aunt.' Joanne Thaine.

151. 'The man standing outside the building was Bill Smith who was sub-postmaster in Layham in the 1920s-30s. His mother and brother lived in the adjoining house. The building ceased to be a post office in 1940, when my parents married and moved into the building as their home.' Nigel Crisp. The posters advertise a whist drive and an enlistment campaign.

152. Coddenham post office (now a private residence) was in a wattle-and-daub building dating back five
hundred years. In the fifteenth century it housed weaving workshops. Standing by his motorbike is
postman Fred Ellis who also appears in Plate 158.

153. The Four Horseshoes public house, Whatfield (now a private dwelling). 'The lady in the photograph is Mrs "Lottie" Thorpe, the wife of the landlord, Ernie. They kept bullocks and pigs and were very hard workers. It was always a lovely pub. Everybody was happy singing and tap dancing on the table. The beer was cheap at two old pence a pint.' Mrs V Smith.

154. The Wheat Sheaf public house at Tattingstone is still a licensed premises. Daniell & Son, a long-established brewery based at West Bergholt, near Colchester, sold its 150 tied houses to Trumans in 1958. The three farm workers are believed to have been employed at Cragpit Farm, opposite the pub.

155. 'The shop at Eyke was added to the front of a much older building about 1900. By the 1920s it was owned by Albert Murrell. His wife and niece assisted him. There was hardly any packaging; everything was stored in tins or large glass bottles. Items were weighed out for each customer. In front of the counter cheaper sweets were displayed to attract children. When I went to the shop with my mother I would get a small bar of Fry's chocolate.' Phyllis Hatcher.

156. This cottage shop was at Grundisburgh. 'My grandfather, Thomas Dimmock, formerly a postman in London, bought the cottage about 1923. My father, Albert, was 10 years old when they moved to Suffolk because of my granddad's weak heart after a bout of rheumatic fever. I loved "helping" in the shop and my granddad was very patient with me.' Ann Ramsey.

157. 'My grandmother, Mrs Mary Pleasance, was the licensee of the Woolpack at Debenham at the time the picture was taken in about 1926. My grandmother is with her hands on my shoulders and with my mother, Mrs Dorothy Abbott, behind her.' Jack Abbott.

158. The Duke's Head public house, fifty yards up Coddenham High Street from Plate 152, is a timber-framed building of great antiquity. It is still a public house of the same name and with the same carved head on the wall. The kitchen to the right of the main building has been demolished along with the structure on which *The People* was being advertised. This was the public well for upper High Street. Presumably capped or filled in, it is now under the pub car park.

159. 'The picture of Clopton Crown shows the landlady Mrs Wilcox with her daughter and my father, Vernon Burch, the driver of the pony and trap. He was delivering meat for his widowed mother, owner of the Grundisburgh butcher's shop.' Keith Burch.

'I have many memories of the Clopton Crown. My father Stanley Wright was licensee from Oct 1937 to 1954. I acted as landlady for three years. Life was quite hard at the Crown. In 1937 [after this photograph was taken] there was no running water, no electricity, wooden floors and tables which had to be scrubbed regularly, coal fires, water heated in a copper. Mild beer was 4 pence per pint, bitter beer was 6 pence per pint, all out of 36-gallon oak casks. These had to be broached by knocking a tap in with a wooden mallet - quite an art to do without spilling. A spigot had to be inserted in the top of the cask to introduce air.

'A packet of 10 Players cigarettes was 6 pence and an ounce of St Julian tobacco was eleven pence halfpenny. A packet of 5 Woodbines was 5 pence. Whisky and gin was sixpence a tot. Bread and cheese was served at 3 pence per portion. Darts, cards, shove ha'penny and bar billiards were played there. There was a steel quoits bed in the back meadow. In 1942 an American Airbase was built at Debach. Many American soldiers were billeted in Clopton and many came to the Crown for refreshments.' Mrs W B Taylor (nee Wright).

160. 'It was such a surprise to realise that these were my grandparents outside the railway crossing cottage in Baylham. Even the dog was there! Their names were Katherine and George Carter. My mother used to live there with them and told how all the cups and saucers slid across the table when a train roared past!' Sylvia Jenkins.

'The gates were opened twice a day for seven days a week for the local dairy herd. Not one accident during all those years, plus the fact the small gates at the sides were never locked, people had more sense in those days.' Mr L Nunn. See also Plate 45.

161. Cotton police station. The baby in the picture, Norman Howling, was born in 1928. He says that the picture shows his father, police constable Percy John Howling, as well as his mother Hilda May, brother Peter and aunt Margaret: 'My father had served in the Machine Gun Corps from 1915 to 1918. On discharge from the army he joined the East Suffolk Police during March 1919, serving in the early years at Leiston. It seems that he must have been transferred to Cotton during 1921, to be in charge of a beat, covering that village plus Bacton, Finningham, Westhorpe and Wyverstone, lodging originally with Mr and Mrs Brooks in Cotton. In May 1921 he purchased a Raleigh cycle from H V Gagger, from the village, it cost £17-2s-6d to provide his transport to serve his area.'

162. The 'ledge' caravan derives its name from the projection along the side of the wagon above the wheels. The wagon doors open outwards (for quick exit), an indicator that this is a Romany wagon. The canvas on the right would have acted as a sunshield.

163. The wagon is more likely to be a showman's wagon than a Romany one. Part of the barely legible plate on the side reads Bury St Edmunds.

164. George Copping, basket-maker and hairdresser of Debenham, is standing outside 35 High Street. Records for the building date back to 1550, though some internal architectural features suggest that it may be even older. No. 37 High Street (the right-hand door in the picture) contains a remarkable sixteenth-century wall painting. The photograph was presumably taken on the same day as the one on page 99 which shows the houses adjacent on the left. The four houses are subdivisions of a building believed once to have been owned by a wealthy brewer.

165. 'This delightful couple were my great grandparents, George and Mary Ann Hill, who lived at Earl
     Stonham. They were the parents of my grandmother Hannah Holden who lived at Poplar Farm,
     Sweffling. George was a horseman and worked, I think, for the Cobbold family. He loved music and
     played the penny whistle. They had eight children.' Pam Rowe (nee Holden).

166. This good-quality timber-framed building has been plastered over for protection. Pentis
     boards are set in above the windows to give protection from the rain.

167. The building's Gambrel (or Mansard) roof is designed to make space for a room on the second floor. The nine-pane windows with a central light are typical Suffolk features. They probably date from the late eighteenth or early nineteenth centuries.

168. These timber-framed buildings were probably built as farm cottages.

169. 'The cottage in the photograph is Rose Cottage in Fishpond Lane, Waldringfield.' G Whyard. The irregular arrangement of the windows and doors suggests that the timber-framed building has been sub-divided and modified.

170. 'The cottage is at Crowfield Street next to the chapel. The couple are Mr and Mrs Scott. The lady on the right is my mother, Mrs Gertrude Radcliffe, who lived next door. I moved there in 1930 when I was four years old.' A Ratcliffe. The patterning on the right of the building has been combed into the plaster in panels.

171. The projection from the building to the right of the women is probably a bread oven.

172. 'The photograph was taken in Haughley, the address then was Stowmarket Road, now it is Fishponds Way. I am in the middle, my sister Kath Kelly (nee Morphew) is on the right. Our friend Jean Eden (nee Cutting) is on the left. The little girl in the front of us was June Burman. The picture was taken around 1928. We were all born in Haughley. I was born in 1915. I think the building was a harness shop years ago.' May Read.

173. Another building with local features and lime plaster, and beautiful eyebrow thatching over the upper window.

174. 'What a lovely surprise I had when I opened today's *East Anglian* to see a wonderful photo of my family at our home in Stratford St Mary, taken I think in the late 1920s. The house, comprised of three lets, was part of the estate of Major and Mrs Hughes who lived in the "Hills" just up the road and our house was named "Ablewhites". I am not quite sure of the date, but 1766 rings a bell as it was on the frontage.

'In the back row, left to right, are my mum, my gran, Elizabeth Norman, and my aunt Nellie. Sitting on the fence is my cousin George Norman. In the front are me, my sister Peggy and my cousin Horace. I am the only survivor. The Norman family were very well known as my grandfather Robert had a green-grocery business and round, which was later taken over by my father Harry, helped on Saturdays by his brother Bill. They did the rounds with a donkey and cart and were a familiar sight, especially in Dedham Street. Dad was quite a favourite of Alfred and Violet Munnings.

'Ablewhites had a huge garden, where we had chicken runs, a stable, pig-sties and an assortment of fruit trees. To the right of the picture was what we called the "copper house" in which was also a large Dutch oven where mum used to cook the Sunday roasts and bake lovely cakes. The third let is at the back of the house to the left. It was surrounded by a cattle yard, stacks and barn buildings, so as children we had a most interesting life.

'Major and Mrs Hughes had plans to modernise the house, but the war came along and later, when the matter was looked into, it appeared the foundations had gone, so we all had to move and Ablewhites was demolished.' Joan Stevenson.

175. Neat front yards in a setting that looks to be more urban than most of the photographs in this section.

176. The wooden bottom plate running to the right of the picture indicates that this is another timber-framed building. There are nice details such as the raised panels on the nineteenth-century door.

177. 'I was amazed to see three generations of my family: my mother Daisy Esme Coppen, my grandmother
Bessie Eliza Coppen, and my great grandfather Joseph Coppen, not forgetting Tiny the dog. The
building was Potash Farm in Bentley, which my grandparents owned and farmed for many years. I was
also born at the farm, as was my daughter, making a span of five generations living at Potash Farm.'
Trish Offord.

178. High summer and an abundance of garden foliage.

179. The notice on the pantiled building is advertising the Bacton and Wyverstone show sports gala on August 3rd, a Saturday. This probably makes the year 1929.

180. This pleasant couple have come to be photographed in their every-day attire: the lady in a pinafore that protects her dress from dust and grime; the man in a collarless shirt.

181. A very simple building, perhaps with single-skin of brickwork, from the late nineteenth century.

182. Lea Cottage, Burgh, a short walk across the fields from Hasketon Hall where it is believed the man in the photograph worked as a horseman. The brickwork, with a Monk bond on the right-hand half, has been applied to an older structure, at least part of which is clay lump. The neat iron gate still stands.

183. Like Layham post office (Plate 151) this urbanised nineteenth-century building shows national rather than local Suffolk features.

184. Timber-framed Nelson Farm, Witnesham used to be called Hogsty Farm. The change of name is presumably connected to the former Nelson public house just across the B1078 in Ashbocking. Perhaps the roof line was raised in the nineteenth century at the same time as the 'Gothick' window features were added.

185. A brick-built building subsequently rendered.

186. The working couple, the man in patched trousers, are standing in front of an interesting timber-framed house. In the right-hand top corner the stack looks to be very old. The dormer window has been tiled – rather than thatched – to give added protection from the elements.

187. A fine example of long-straw thatch. Features such as the garden gate show the care people took over their properties.

188. Mill House, Grundisburgh, around 1928, was the home of Robert Charles and Emily Nunn. Robert owned and operated the postmill which stood close to the house. The base remains as a private residence. The children are (left to right) Alice, Eric and Eileen Nunn, three of the five children raised by Robert and Emily. Eric was to be killed in the Second World War.

189. Unlike most of the Titshall pictures, this appears to be of a back yard rather than the front. In the
foreground the hutch contains fowl. The timber-cladding - with feather-edged boarding - is more
common in Essex than Suffolk. The building to the left could be an added-on wash-house.

Names of Places

Other titles from
Old Pond Publishing

The companion to Just a Moment
In a Long Day
The first collection of Titshall photographs emphasised the rural scene. Half the plates show working horses and their horsemen, and there are substantial sections on steam-powered threshing and rural trades.

Novels by H W Freeman

Chaffinch's
A moving novel that depicts the life of farm worker Joss Elvin and his struggle to raise a family on 19 acres of Suffolk farmland.

Down in the Valley
In the 1920s young Everard Mulliver leaves the town to settle in a quiet Suffolk village. He learns to fit into the countryside and acquire a deep-rooted passion for the land.

Joseph and His Brethren
H W Freeman's first novel follows the story of a Suffolk farming family through two generations in the late nineteenth century.

Hugh Barrett's sharply detailed memories

Early to Rise
Hugh Barrett's vivid account of life on a Suffolk farm in the 1930s. Living in as a farm pupil, Hugh learned to plough, build a stack, hoe beet and grind the pig food.

A Good Living
Managing farms from 1938 to 1949, Hugh Barrett encountered a range of characters from gentlemen farmers to ex-miners on land settlement holdings, with wartime profiteers and downright rogues for good measure.

The Land Army

Land Girls at the Old Rectory
An entertaining account by Irene Grimwood of what it was like for a town girl to join the Land Army in 1939-45. She and her lively friends learned to hoe, build stacks and cope with livestock as well as American servicemen.

A Land Girl's War
Joan Snelling became a tractor driver during her wartime service in Norfolk. Her book recalls the dangers and tragedies of the period as well as its lighter side and her romance with an RAF pilot.

The west of England

Charismatic Cows and Beefcake Bulls
Sonia Kurta's stories of farm work are mostly set in Cornwall in the 1940s where she observed with amusement the behaviour of her colleagues and livestock.

Farmer's Boy
Michael Hawker's detailed recollections of work on north Devon farms in the 1940s and 50s includes sections on corn and potato harvesting, horse power and early tractors.

DVDs about farming life
Old Pond Publishing has a growing collection of DVDs about farming life including the Anglia 'Bygones' series and programmes featuring Paul Heiney.

Free complete catalogue:

Old Pond Publishing Ltd, Dencora Business Centre, 36 White House Road, Ipswich IP1 5LT, United Kingdom

Website: www.oldpond.com

Phone 01473 238200   Fax 01473 238201   Email enquiries@oldpond.com

Acknowledgements

All the photographs in this book have been printed from the archive of the Titshall brothers' glass negatives saved by Doug Cotton and owned by David Kindred.

The interpretations of the photographs are conjectural because we have no written record from the photographers themselves. The first source of information has been readers' responses after photographs were published in recent years in the *East Anglian Daily Times* or the Ipswich *Evening Star.* These responses have been filed by David Kindred. Some are reproduced here in edited versions, in quotation marks, with the names of the respondents. Others have been adapted and have not been specifically credited. We are grateful to all those who did make a response.

I have consulted experts on specific subjects and am very grateful for the time they gave up and for their information:

Graham Austin (railways)

Ken Austin (railways and steam)

Brian and Ivy Bell

David Chaplin (Suffolk Horse Society)

Brian Dyes (Ipswich Transport Museum)

Jill Freestone (railways and Stoke)

Stuart Gibbard (*Old Tractor* magazine)

Frank Grace

Miriam Harrup (Co-operative Society)

Ken Leighton (railways)

Bill Marland (Suffolk Horse Society)

Michael Munt (historic buildings)

Russell Nunn (Ipswich Transport Society)

Derek Rayner (Road Roller Association)

Richard Smith (Ipswich Dock)

Nigel Stennett-Cox (motorised transport).

Thanks are due to Lesley Smith for proof-reading. Errors, misinterpretations and omissions are entirely my responsibility. David Kindred and I would be very glad to receive further information about the photographs (which can be sent to me at Old Pond Publishing).

Roger Smith
2007